Instant Profits with Instagram

ISSA ASAD

Copyright/Disclaimer

CONTENTS

CHAPTER 1:
SET UP YOUR INSTAGRAM ACCOUNT

Today, businesses like individuals can use image sharing sites such as Instagram to help grow their business. In fact, images can help enhance your communication with consumers and the experience they have with your business, and it can help your business grow when images are properly used.

A site like Instagram is extremely easy to start up on, and allows you to connect with your customers in a manner you have never been able to do in the past.

Signing Up To Instagram

In order to connect to Instagram you simply have to set up an account with the site. You will sign up using the same handle that you use for your Twitter account. This is not only going to make it easier for your customers to find you, but will also be easier for you to remember your user names for the different sites you are registered with.

A profile picture should also be added; this can range from your company logo, to new product offering, or any other image that will inform customers who you are. Additionally a site link should be added. This allows consumers and new prospective visitors to visit your online site, which will help grow your business. You can also connect to Facebook if you are on the social media site to give your current followers one additional place to find you on the web.

Once your account is set up, you can begin adding images. Of course you have to abide by community guidelines, which are all laid out for you when you register. But adding images of new product lines, future product releases or even images your clients send to you will enhance your online presence as a business. It will also allow your customers to connect with you in a visual manner rather than simply reading about what you are up to, and learn what new products are in the works for your business.

How Do You Use Instagram For Business?

Using the online photo sharing site for business versus personal uses will vary. For instance, the images you post, the frequency of posts, the messages attached to the images and possible images you will share from other accounts will vary. For this reason, you have to understand the business uses of Instagram in order to ensure you are utilizing the site to your advantage. This is opposed to accidentally turning your customers off by offensive posts, or posts they would not expect to see from a business which they trust and make purchases from.

The Instagram Business Blog

The Instagram for business blog is one that has taken off in recent months. Due to the influx of businesses joining the social media photo sharing site, Instagram responded by providing businesses with a tutorial of sorts in order for them to understand the site uses and how best to utilize their accounts.

The blog provides:

- Tips for photo sharing, and when to post photos for customers to view.

- Brand spotlights, and how to incorporate them in to your account and photo sharing page.

- API examples, so businesses understand how to properly post them.

- News from Instagram Headquarters.

This information and other relevant business information will not only help businesses take off on the social media site, but will also provide business owners the in-depth knowledge required to build their brand on the site. The more informed businesses are and the more they understand how to utilize the tools which are in front of them, the easier it is for them to build on their brand, their product line and to eventually bring in new customers to their site and business.

Keep Up With the Instagram Business Blog

It is important for business owners to keep up with the Instagram business blog. In addition to information on setting up an account and how to properly use the photo sharing site, it provides tips and information as it is released.

From new releases and updates to the site, to the new terms and conditions, to relevant information for any particular line of business. The businesses which do keep up with the blog and who continually check up on the blog, are going to get ahead on the photo sharing site.

Business are eventually going to build in it by incorporating video and other images which are relevant to their consumer base.

Create a Profile That Works

Optimizing your Instagram profile is essential to visitors being able to find your site. Not only will a well-developed profile page call their attention, it will also grab the potential site visitors who may be interested in your site, but have never clicked on an image or have never visited your site in the past. So, in addition to keeping your current customers happy it will also help draw attention to the potential customer that has never visited your site in the past.

Just like Twitter and Facebook, your Instagram profile must include a brand name so that your consumer will know how to find you on the social media platform. The brand name will be limited either by the number of characters which you can use to create the headline, image size or other attention to branding which are laid out by the guidelines of Instagram.

"How to Build the Perfect Instagram Profile" by Gerry Moran. For the original size image, go to http://marketingthink.com/infographic-to-build-the-perfect-instagram-profile.

In the above diagram, not only do you learn how to incorporate your brand name, but also how to use it in order to call attention to visitors. From the location of the wording, to location and image attachment, there is a perfect art to developing your Instagram profile page so that it really stands out.

Geo-tagging, using the right profile picture, properly hyper-linking your account and more is necessary. Although this information may not seem relevant to you when setting up your account, it can make a world of difference to the visitors who are looking for your account or who want to learn more about your business. So properly setting up the site is essential when you are building your account and are coming up with the information to add to the account profile page.

What to Include In Your Profile Page

When building your profile page, it may be difficult to determine what is relevant, what you should include and what you should exclude from your page. The right information and the things that call attention to your visitors, are those which you should include on your profile page.

Some things to consider including are:

- A brand image. Not only will site visitors be able to determine it is truly your account, they will also be able to find exactly what they are looking for when they see your brand image.

- Profile picture. The profile picture should encompass what the business does, what the business stands for and what customers can expect when they choose to do business with the organization.

- Notifications section. This allows followers to find your site from anywhere and allows them to receive information about when something new is posted, or new images are being uploaded to your Instagram account page.

- Site URL. Your site's URL must be clearly visible, as should any hyperlinks which you choose to add to the site. If a visitor does not know where to find you outside of Instagram, how are you going to grow your business by simply posting images to the site?

These are some of the most important pieces of information that you want to include when you are building your profile page. It is basically a blueprint of what your business does, what you offer to your customer and what they can expect from you in terms of product offerings, or the services which you sell as a business owner.

Of course adding the right images to your account is extremely important, but if your account profile is not properly set up it can greatly reduce the number of visitors, the number of followers and the number of people who would potentially see the site that you are building.

Knowing the Limits

There are certain things which can't be added to your profile, or other restrictions which are in place when you are developing the profile page. This can range in things such as the number of characters you can use for your headline, to the size of an image you can use on the main page of your profile.

Understanding the restrictions, knowing what limits are in place and understanding what you can and can't do on the site, are a few things to keep in mind as well as you are developing your page. It will not only eliminate your Instagram account from being flagged, and eventually deleted, it will also make setting up the page much easier for you as well. So, understanding how to optimize your profile, all while understanding the limitations that are in place, is something that you must understand as a business owner who is setting up your account for the first time.

Doing Business with You

In concluding your profile, you should include all of the relevant information a customer might need in order to get in contact with you. Suggestions include a company phone number (or numbers), information about the CEO and site owner, a physical address and of course all URL and online contact information. Email addresses which customers can reach you at, or even a cell phone number, are pieces of information that you should consider adding to your profile page as well.

The more ways you give your customers to reach you, the more likely it is they are going to contact you, to visit your online site or to visit your physical store if they live in a region where you operate your stores. Like building a website, if a customer does not know how to reach you, how are they going to do business with you?

The same goes for Instagram. Providing the relevant information required to get in touch with you, or to simply ask a question about your business is the only way for customers to find you and for new, prospective clients to know you exist somewhere outside of the online spectrum.

Setting up your account is quite simple. Following the Instagram guidelines, knowing what to add and what to leave off your profile and knowing how to quickly and easily set up a page that stands out are some of the things you need to know as a business owner. With your new site in place you are ready to grow your business, gain new followers and reach a new prospective audience for your business as well.

CHAPTER 2:
ALWAYS TELL A STORY

"A picture is worth a thousand words."

You have probably heard this saying a thousand times, but it is true. Using pictures to relate to your consumer and to prospective consumers is an easy way to connect. For example, an image of a new product you are going to sell is far more descriptive than a paragraph you would write about it.

People can see the color, the fabric, the texture, the details and so forth. Although words can convey certain details about a product, they will only go so far in doing so. For this reason, when you are deciding which images to add to your Instagram account you want to make sure you are always telling a story. This is the opposite of simply adding random pictures which have nothing to do with your business, or the information which customers would like to collect from the images that you choose to share on the social media platform.

Take Advantage of the Space

With Instagram you have a whole new way to tell a story - through images. So it is important to do so. You want to make your brand stands out and you want people to connect with your company on a human level.

By adding both fun (or personal) photos as well as business photos, you are going to be able to do this. But again, there must be some type of connection between the images so that the visitor is going to be able to put the story together in the end. If you are simply adding images on a random basis or throw in images that are not connected in some way, it is going to take away from your story.

Telling the Story

There are many ways to use the social photo sharing platform to share your story.

Some photos you may choose to include are:

- Photos of your product lines. Whether you sell clothing, linens or a particular service, let customers see it. When they get a glimpse at a new product or the product in design, they are going to feel connected to you and your business.

- Images of your employees. Even if a customer never sees the people behind the scenes (because your business is solely online), having a face behind the voice (or in this case computer screen) is always nice. The people who work for you and develop your brand are a part of the story you want to tell.

- Photos of events or charitable causes. If you can connect with your clients on a human level it is a great advantage for you as a business owner. So consider adding the images that show charitable work you do, people you help in the community or fun things you do with locals. These are all a part of your business and can be a part of the story that you tell your clients and site visitors.

The more you can connect with people and the more you can connect to those who are seeing these photos you add over time, the easier it is going to be for you to build your brand.

You want your Instagram account to tell people who are seeing the pictures something about your company. Any of these types of images are going to do just that. Not only will they allow your customers to see what you do and what products you have in development, but will also give them a glimpse of how your company relates to them in other ways, apart from the purchases that they are making from your business as well.

Strike a Balance

Of course depending on the type of business you run, the type of product you sell, the service you offer and the type of clients you are dealing with, the images you choose to share will vary when you are developing the story you are telling.

For example, a business in the line of finance is not going to share as many fun images or goofy pictures, as a company that sells children's clothing might share.

Understanding how to strike a balance between the fun and the business photos that you upload is extremely important when you are adding the images to your story. Also, placement of the pictures, frequency of certain types of images and images that you choose to share (if it was sent to you by a customer), are all factors to consider when you are adding images to Instagram, and adding them in any particular order to the site.

Whether you are adding one photo each day, or twenty photos each day, the right mix of photos are going to complete your story and are going to strike a balance between the business you do, the product or service you sell and the type of clients that you engage with as a business.

Use Statistics

There is a like button for a reason on Instagram. Basically these two buttons are going to be your guidance as to which type of image you want to share.

If customers like an image, they can tell you this and even add a comment as to why they like it; the same goes if they do not like an image. If certain types of images you are posting receive a higher percentage of likes versus dislikes, this should indicate something to you. The same goes on the flip side; if customers are disliking the images you share or seem to be upset or otherwise insulted by certain images, you have to take immediate action to remedy the situation.

Instagram will allow you to instantly remove an image. This is one way to counter the dislikes you are receiving and it is a great way to show your customers that you care about their opinion. Not only are you removing images that are offensive or that they otherwise do not want to see, but you are acting quickly so as to avoid losing such customers and site visitors who do not like certain images you are using.

It is important to read into the likes and dislikes for images, as well as the comments which are being left behind by followers who choose to comment on the images you post. This information is not only going to help you determine what parts of your story customers and visitors enjoy, but will also allow you to modify your story in the future so that you remove the chapters which are not relevant to the customer who is visiting your account page.

Striking a balance is essential to developing your account; but, if customers are continually telling you they do not want to see a certain type of image, then you want to make the modifications as needed. Sometimes a particular type of visitor is only going to want to see business images, or the fun images.

Either way you have to use the statistics, use the comments and listen to what your visitors are saying so as to ensure you do not lose any followers on the social media platform. This will also ensure visitors are getting the most out of the images that you choose to share with them as a business owner.

Engagement of Images

This generally comes in the form of comments that are left behind on the images you post by people who click on a photo and like it. As a business owner (or as the person in charge of adding images to the Instagram account), you have to track this information on a regular basis.

If possible, it is a good idea to check the comments and feedback which is being left behind on a day to day basis. Doing this allows you to see which photos your visitors most appreciate, which ones are getting the most traffic and which ones are leaving the biggest impact on them, both positive and negative impact.

Because not every person is going to react to a particular image in the same manner, reading comments is just as important (if not more important) than looking at the likes for different images. Just because two people like the same picture or dislike the same image, it does not mean it is for the same reason. When you read the comments, you can learn more about what visitors have to say about the images you are posting and you can pick up more about what they feel when they see certain images.

This engagement and comment sharing also begins discussions about your product and your company. So reading the comments, the interaction between your followers and what they are most talking about in each image is a great way to learn about what they like and what they do not like about different images which are posted.

This gives you an advantage by allowing you to make changes as needed and properly modify your story on the image sharing site, and to add the images which are most interesting and which are most liked by the followers and visitors that are coming on to your page.

Do What People Like Best

No matter what, it is impossible to please everyone. For this reason, you have to find what works best. Whether it is striking a balance between fun and business, or simply sticking to one side of the spectrum, your Instagram photos have to tell a story and they have to engage the visitors who are going to be reading the story that you are sharing with them.

Since there is no way to please everyone, you have to find what works for the masses, and this is what you should continue to deliver to your site visitors in terms of the pictures that you choose to share with them as a business.

Find out what your followers like best, and give this to them. By doing this you are not only going to be gaining new followers, you are also going to realize your business is going to grow. As your followers are going to share your account page, new followers will eventually visit your site as well.

So remember to tell a story, but also to make modifications and changes along the way as needed. You also have to remember that you can't please every person that visits your page and for this reason should make use of the information you do have, to learn what visitors like best and to give them this when posting new images to your account page.

CHAPTER 3:
MARKET YOUR BRAND USING TRENDY AND INDUSTRY-RELATED HASHTAGS

Hashtags are one of the biggest crazes on Twitter and on Instagram. If a celebrity uses a particular hashtag it is sure to blow up in an instant. Or as a business owner, you can use hashtags to indicate a sale or other new product release. If a customer uses the hashtag, you can give them a discount on a future purchase or other such incentive for using it.

Using the current trends and incorporating the trends to your account and photos you share is likely going to gain more attention. Not only are more visitors going to look at images which have a catchy headline and hashtags, they are also more likely to share these with friends and possibly others who do not follow your Instagram account.

In turn, this means new followers, it means more likes on the images you share and it potentially means an increase in sales figures for a particular product which has a catchy hashtag. All this comes by simply posting a photo on the social media site and properly utilizing the trends which are currently well known by the visitors and followers who follow you on the social media platform.

Use the Latest Trends to Market Your Brand

Hashtags allow Instagram users to connect and join in other stories on the social media platform. Certain hashtags have withstood the test of time, such as "throwback Thursday."

As a business owner, it is a good idea to develop a hashtag which allows your followers to get in the mix and to share their own photos, as opposed to simply adding your own images to the site. For example, a hashtag asking customers to share a selfie with a new clothing line they purchased (which was just released) is a great way to develop on your story.

When a follower posts their own image with the hashtag (#), it is not only going to allow other visitors to your account to view the images, it will also allow you to connect with your customer.

When the customer knows there is a person behind the computer, it is more likely to make them feel closer to the business and to your brand. The closer they feel to your company, the more likely it is they will make future purchases, and the more likely it is they are going to share the product line or services that your company offers with their friends and family.
The hashtag is even a great way for other people on Instagram to learn about your business. If one of your followers is using one of your hashtags, if a friend of theirs (or someone who follows their account) sees the hashtag, they are likely going to follow your business account.

So, not only do the trends such as hashtags allow you to connect with your customers on a human level and allow you to engage by commenting on their photos, it also allows you to build on the number of followers you are going to gain because of a cool hashtag. The more attention you can garner, and the more free marketing you can gain from your followers (without exposing them), the more people are going to see your brand name, the more they are going to learn about your products and the easier it is going to be for you to grow your business online.

All of this is going to come as free publicity for you, and is a great way for you to engage with your followers through the photo sharing platform online, and for you to gain more attention for the new merchandise, or a sale that you want to share online.

Use What's Relevant

If you can find a hashtag that is relevant to your business (such as #SelfieSunday), this is all the better for your business. Due to the fact that these hashtags are already developed, it will allow others who are on Instagram to see the hashtag being used by your follower.

This is then going to garner more attention to a particular product, to new merchandise you are selling or to other relevant information about your product. If it is a popular hashtag and if your followers are participating using images of your merchandise, this is free marketing for you as a business owner, and it is a great way to spread the word about your business and the product line that you offer for sale to your customer base.

Keep It in the Industry

There are a number of industry related hashtags which have also become popular on Instagram over the years. It is important to take advantage of these as well if you are in that particular industry. Attaching the hashtags to images you post on your account page makes it easier for people to find them.

Some ways to incorporate hashtags are:

- Events. If you are hosting a function or a formal event, using the industry related hashtag to showcase a new product, or something that your business does, is a great way for site coordinators to view the image and share it on their site.

- Customers. If a customer is using one of your products and you can use an industry related hashtag, do it. Not only will that particular customer share the image on their account, but will possibly share it on other social media platforms as well.

One of the main benefits of using the industry related hashtag is the fact that it is going to be re-posted by others who are tagged, and it is also going to be shared through other social media platforms. If attendees to the event use Twitter or Facebook, they are likely to post the Instagram image on these sites as well; the same goes for the customers who re-post the image.

So you are going to gain more visibility on Instagram as well as on the other social media, and social sharing platforms out there as well. From there, others are going to find themselves, will see an image a friend posted or will see a hashtag, and will also share these on their social media platforms.

Basically, the industry related hashtag is going to garner more attention for the product, or for a particular brand you are trying to gain more attention for and people are going to share it for you.

It is important to use these as often as possible, and to tag as many people as possible when you post them on your business account. Doing this allows more followers to see they were tagged, and this will potentially lead to friends being tagged and others who are on the social sharing platform to sharing the image as well. More people are going to see your product or your brand name because of this, plus this comes to you as a free form of marketing as you do not have to pay for the images that you are sharing, and you do not have to pay your followers for sharing the images that they are sharing for you.

Track Your Hashtags

To make sure the hashtags are having a positive effect on your business, it is important to track them.

One such place that you can do this is at: http://totems.co/

Not only can you boost your recognition online using this site, you can also track the relevance of your hashtags, how often they are being shared and how much you are getting out of using them.

Some information that the site will provide you with includes:

- Contributors and who is sharing the images which are posted using hashtags.

- The content and how it is being discussed online.

- Engagement, and how many people are sharing or re-posting the image with certain hashtags.

- Context, and how it relates to the images you are sharing.

Basically the site is going to give you a breakdown as to how the images are being received by those who follow you on Instagram. It will also provide you with the necessary feedback so that you can determine whether or not certain hashtags are working, which ones are working and which changes you have to make.

Tracking the relevance of different hashtags you are using will allow you to determine which ones work, which ones don't and which ones you may want to try in the future. It will also allow you to make the necessary adjustments as needed, or to make changes to the types of images you are sharing as needed. All of this can be determined by using the appropriate sites and seeing what kind of an impact different images are having. Lastly, it will also show how often the hashtags are being shared, or re-posted by your followers on the social media platform.

Know Your Industry

Simply because a particular hashtag is industry related, does not mean it is going to help your business out.

As a business owner, you need to use analytics and you have to use the information that is presented to you to determine if these hashtags are helping or hurting your business. Simply because something is trendy (such as the #), does not mean it is right for your business.

By tracking this information, you are not only going to know when to use it, and when to leave it out, you are also going to know how to engage your followers to use the hashtag to ensure you are getting the most out of it.

More often than not, people are going to respond to the use of hashtags; it is simply a matter of finding out which ones work, what to use when to use them, and what to avoid posting. When you do this, not only are you going to see a major increase in the frequency at which your images are shared, but you are also going to realize a difference in the number of site visitors that are following you on Instagram as well.

So, learning the ins and outs of how hashtags work, using the appropriate ones and knowing how to use them, are some things to keep in mind in adding the hashtag to an image. Proper use will go a long way; but overuse, or improper use is going to hinder your growth, and may eventually lead to certain followers turning away from your business if they feel they are being exposed in any way.

CHAPTER 4:
STIMULATE YOUR CUSTOMERS' EMOTIONS

Relevancy is the most important factor to keep in mind when you are posting photos to your Instagram business account. The images you choose to share should not only be relevant to your brand, the products you sell or the service you sell, but should also be relevant to your customers as well.

It should elicit some type of emotion; whether it is to encourage them to partake in a movement, to try something new or simply to go out and purchase your product, the images you share have to call out to your customers in one way or another.

Make It Inspirational

The images you share should be inspirational. If your business sells health products, add images showing athletes using the product. If you are a company that helps people learn how to cook, add images which show how to incorporate ingredients, or show others cooking a dish they would love to learn how to cook.

Whole Foods Market by Whole Foods Market Instagram

Regardless of the type of business you are in, what you sell or what you want to encourage your customer to buy, it has to be inspirational in some manner, and it has to speak to them at a deeper level if you want them to react to the images you are sharing on the site.

The above picture is an example of a great Instagram business account. The mass merchant sells health foods, promotes people to engage in a healthier lifestyle and to choose healthier foods to cook with. Not only does it showcase some of the products which are sold, but also includes images of people going out and being active or creating new, healthy recipes to try for their family. It encourages their visitors to get out there and do something about their health, and inspires their followers by showing them what others have been able to do with the help of their business.

In a similar manner, you have to engage your followers and you have to inspire them with the images you post. If the images are dull, do not speak to them at a deeper level and do not motivate them to make changes, then they are never going to want to purchase from your company.

Using the right images, focusing on images which relate to your business and also relating to your customer in some way is extremely important when deciding on the images you are going to share on your site.

Inspire Your Visitor

A great way to sell to your customer is to inspire them. When adding images to your business page, use those which will inspire your customer to do the same. If you can add images of previous clients who have lost weight (for a personal trainer site), or can show a client learning how to cook (for a business that teaches people how to cook), then this is going to resonate with your followers.

It is going to show them people who were in a similar position as their own, making positive changes and doing things that bettered their lives. It will in turn inspire your followers to want to get up and do the same, and will make them want to work with your business to help them make the positive changes which they seek in their own lives.

Showcase Your Employees

Especially if you have a business that is run strictly online, your customer never gets to see a face behind the product or work being done. A great way to showcase the great people who work for you, is to add images of them to your Instagram page. It not only shows your followers you are a real company, with real employees, it gives your employees thanks where thanks is due.

You can add images of employees making a product or teaching others a service that you offer for sale. You can add images of employees at a work function or event or at a training seminar they have to attend. It is a good way to celebrate your staff and show them you recognize their hard work. It is also a great way for your customers and followers to really see the people who are behind your products, and behind the merchandise that they are buying.

Adding images of employees also gives your business more credibility. When your followers can see there is a physical building with people doing the work, it is less likely they are going to feel they are being taken advantage of. Not only do they see the product being made, they see the people who are behind it as well. It will encourage them to purchase and will encourage your customers to want to make future purchases, and continue to do business with your company in the future as well.

Acknowledge Your Staff

If a staff member has a particular skill and is the only one in the company that can do something for your business, acknowledge it through images you share. Or if they reach a milestone, such as working for your company for several years, make sure to share these images as well. Not only will it make that staff member feel as if they are a part of the family, and are truly a valued employee within the organization, it is also going to show your followers you acknowledge and care about the people who are working to make the product lines they sell.

Doing this will give staff more incentive to carry on and continually perform great work for your business, and will show them they are an integral part of your business. It shows your followers you do care for your staff and that it is not only about the end product, but also about the work that goes in to making the product lines they are purchasing from you.

It is a win-win situation for you as a business owner to share these images and is a great way to help grow your story that you are telling on Instagram. You are now incorporating a family, an inspiration and something that brings your employees, your business and your clients together on the site.

Show You Are Human

Instagram is the perfect platform for you to share the fun that takes place behind the scenes as well. Since you have chosen to integrate images of your employees, add some images of what takes place behind the scenes as well.

Post images of mistakes that are made, fun times which are had in the office or other fun and play that goes on behind the scenes. No matter where you work, it is not work all the time; there has to be a little fun in any work setting to offset the difficulties of the work being performed. Instagram is a great place to post these images and to show your followers you are a company that does focus on the work and end product, but also knows how to let your employees have fun when they are doing their work as well.

By posting these images, it shows you have a human side as well and allows you to connect to your followers on a human level. Since you are most likely never going to see or speak to your followers, it allows them to see a face behind your name and allows them to see there is a caring and fun human behind all of the work that is being done within your organization.

It shows that your company does not only care about work, but also cares about having fun, about your employees' happiness and about delivering the best product to the client without taking things too seriously all of the time. Just because you are at work, does not mean you can't enjoy what you are doing or what you are delivering to your customers. By posting these fun images and the hijinks that take place behind closed doors, you can show this to your followers and anyone who sees the photos that are posted on your Instagram page.

Your business should not be all work and no play. Make sure you show this to your followers and let them see some of the fun times your employees do have at work. It will make them feel more connected to your business.

Bring It All Together

Inspiring your customers to get up and act, connecting to them on a human level and bringing them closer to the employees and the people who are actually behind the products you sell to them, are all things to consider when you are deciding which images you are going to post on your Instagram account page.

You have to strike the right balance, and you have to know when enough is enough of anything (whether it is the inspirational images, or the fun that takes place behind the scenes). But you do want to blend it all in, and you want to make sure your visitors and your followers see a little bit of everything when they visit your Instagram account.

If you are solely interested in profit margins, making the sale and do not show concern for your customers or those who actually do the work for your business, it is going to draw people away and is going to show the consumer that you do not really care about their wellbeing or their personal needs. Like anything else, there has to be a balance when it comes to the work. You have to show the customers who are purchasing from you that there is a purpose behind the work you do, and that you do care about their happiness with the products they are purchasing from you. By sharing these images on your account page, you can easily show your followers they are a priority and that it is not simply about making a sale to them.

Plus sharing these inspirational images, the fun at work and the different images which relate to your customers is going to add a little bit of flair to your account. It will help strike the balance between fun and work and help you connect to your followers on a deeper, more meaningful level as well.

CHAPTER 5:
ENHANCE YOUR INSTAGRAM PICTURES WITH FILTERS AND APPS

Apps are another one of the major Instagram trends today. Like hashtags, they allow people to share images in a number of unique ways. Since there are so many apps available for nearly every platform out there, apps are going to allow your followers to view your images, and to share them with others in a number of unique ways.

Using the best apps for photo sharing, and adding in the right filter to ensure your images always look top notch are a couple of things that you should do as you are developing your account, and as you are adding in the new images that you want to share with your followers on your Instagram account page.

Apps: The Power in Social Sharing

You can easily enhance your photo sharing as a business owner on Instagram by letting your followers view the images you post through the many apps that are available for photo sharing today. There are so many apps out there that it allows your followers to find and share images in a number of ways.

Some apps which are available today allow your followers to:

- Print out the images which are shared, and have a physical copy. This not only allows them to share the images online, but also in print form as well.

- Search for images. Using a specific keyword, hashtag or other relevant phrase, your followers will be able to find a specific style of image that you have posted on your Instagram account. This not only eliminates the need to go photo by photo to find what they are looking for, it also allows them to group together a specific type of photo that they would like to view, and potentially share with others in the future.

- Use email to subscribe. If an individual who is on Instagram finds a photo that they like posted by your business, they can view others. Certain apps then allow them to follow your business via email and subscribe to your Instagram page, which means new customers and more people seeing the images which are posted on your account.

- Use folders to download images. Certain apps allow your followers to download all photos, and create archives and folders to place the images in. This allows for more organization, and allows your followers to view, share and post the images you have on your account to other social media platforms as well.

By simply incorporating these apps to your marketing mix, you can quickly and easily share more photos, get new followers and potentially gain new customers, which will in turn help increase your profit margins.

Add the Right Apps to the Mix

These, and other popular photo sharing apps should be added to your account so your followers can use them. The easier it is for your followers to find an image they want to see, to subscribe to your page or to share an image, the more it is going to do for your business in the future.

So when adding images to your account, make sure you are using the right apps, which make it easier for your followers to view the images in the way they want to view them and to easily share the images which they like best.

There are dozens of apps available for Instagram; so you do not have to limit yourself, nor do you have to choose between similar apps which will allow you to set up your account in a particular way. The more you use, the more structure you give to your Instagram page; the easier you make it for your followers to view what they want to view, the more beneficial it is going to be to your business.

Utilize these tools, as they are not only simple to use, but a majority of them are free for you to use when you are setting up your page. So you can quickly and easily create folders, or move images to a specific section of the page when you are adding images to your account. In turn, it allows your followers to see what they want to see, to easily share images and to view the images that you post to your Instagram account, on any platform they are using to access your business page.

Create the Best Fit

When using Instagram, there are also a number of filters which you can use in order to ensure each image you post looks as professional (or amateur) as you want it to appear. If you want to highlight a color on a piece of fabric, there is a filter for that. Or, if you want the image to look as natural as possible as if it was not modified in any way, there is also an app for that.

Due to the fact that you are setting up a business account on Instagram, it is extremely important that you only post images that look great, are not grainy and allow your followers to see your company in the best light. For these reasons, it is important to make use of the filters, juxtaposing positions and other tools that are available to you when you are posting an image to Instagram; this ensures your customer is always going to see that image, the way it was truly meant for them to see it.

Most top companies are on Instagram today (nearly 59% of top businesses), this means, regardless of the industry your business is in, or the type of product you sell, your competition is likely using Instagram.

Therefore, you need to make sure the images you are posting not only look better than theirs, but also showcase to your followers why they should purchase from you, and continue to do business with you. This is opposed to turning to a different company that offers a similar type of product or service, which is in the same industry as your business.

Although filters do modify the appearance of certain images, the color and other design elements, they are not simply about the aesthetic appeal when it comes to sharing images on Instagram. The filters which are chosen by different businesses can also say quite a bit about the business, their ethics, the work they do and the messages which they want to convey to their customers.

So it is important to keep this in mind when deciding on the right use of filters, and deciding how to present the images that you share on the social media platform. Not only does a certain type of lighting let followers view an image in a particular way, it will also give a certain view about your business. It is important to know what the different filters say about your business, and how you conduct business, and also to know how this is going to be viewed by those who are following you on the social media platform.

You want your business to be viewed in the best possible light in terms of your honesty, and how you conduct business with your clients. For this reason when posting images and choosing filters, juxtaposition, images, direction and positioning, you have to know how to properly do so, and how the choices you make are going to affect your audience who will be viewing the images that you are sharing on the site.

Not only do you want it to be a great looking image that you post, but you also want it to say the right things about your business, and how you conduct business with the customers that are following you on Instagram.

Don't Remain Stagnant

Since the filters, juxtaposition and other factors tell your followers a little bit about you, don't keep things the same with every single image you post. If you choose to go with a lighter filter for one image, choose a different filter for a few other images that you post. This not only keeps your followers guessing, it is also going to show your followers that you know how to use these tools on Instagram, and that you want to present them with the best possible view of the images that you are sharing.

When you shake things up, when you present different images in different ways. When you are using all of the available resources you have, this is going to have an impact in how your followers view you and what they think about the images that you choose to share, about your product and about the business that you operate.

It is a good idea to use natural filters as well as those which highlight certain product attributes. It is also important to properly place items that are being photographed to ensure the right focal point is seen with each image posted on your account. A filter change can make a world of difference in how a product is viewed, and whether it is going to receive more likes or dislikes when you finally post it to your Instagram account.

You want your images to look good, and you want your followers to be able to see the main focal point with each image you choose to share on Instagram. This is where filters and proper positioning will come in to play. Use the tools that are available to you and learn how to share images in the best light.

Make sure you keep this in mind, utilize the tools properly and make sure you understand how using filters and other tools is going to showcase more about you and your business.

Furthermore, learn how to share on different platforms. Apps are one of your best friends today—use them. Not only will they allow you to share your images in the way your visitors want to see them, but will allow you to share them on different platforms. People have more options in terms of what they are seeing and what they can share.

Apps will make image sharing a breeze, will allow your visitors to view images how they want to see them and will eventually lead to more sharing and subscriptions.

CHAPTER 6:
NETWORK YOUR BRAND ON INSTAGRAM

Networking your brand online is not as easy as one might expect. It is important to cultivate the right following, to get your brand name out there and recognized by the right people and to market it to those who are actually interested in what you have to offer for sale.

Getting more followers on Instagram is one such way to grow your business, grow your brand and to eventually earn higher profit margins as a business owner. It is important to know how to grow on Instagram, how to gain new followers and how to really share the product offerings you have to those who are most interested in it.

Gaining Followers

Gaining new followers on Instagram is not as easy as asking people to follow you; but it is not a great mystery either. One simple way to gain more followers is to connect to other social media platforms.

Connect your Instagram account to your Facebook page. This will automatically make people connect and follow you. Not only will it ask them to follow you on their timeline, it will post new images, updates and inform people when new images are shared on your Instagram account. Facebook is still one of the most (if not the most) popular business social media platform; utilize it, and connect your accounts to let your current followers you also have an Instagram account. You can even ask them to follow you through Facebook.

Engage and follow other people to gain new followers. This is also an easy way to gain new followers. Some people who see you are viewing their images and post comments on them will follow you back, even if they are not interested; others will wait for you to request to follow you back. And if you are following other businesses, this is a great way to gain new followers. If they follow you back, many of their followers are automatically going to follow you back as well.

The use of popular hashtags is another easy way to gain new followers; but you have to use those which are relevant to your business if you want people to follow you. If you are using hashtags related to one industry when you are in a completely different industry, it is going to upset people and they will immediately unfollow you.

Using relevant hashtags is a way for people to see the images you post, even if they are not following you. From there, they can view more images and if they find your account to be of interest or something that they are looking for, it will more often than not lead to them following you.

Even if new followers do not lead to more sales, it does lead to more recognition and the more people seeing images of your products, the more people will be learning about your business online. This is always a good thing.

So, get yourself out there, engage, follow and connect. Doing these three things will greatly result in an increase in the number of followers you see, and the number of visitors you will see coming to your business site over time as well.

Networking Is Critical

As laid out above, gaining new followers does not really require some difficult to understand formula or algorithm; common sense would dictate that doing these things will lead to more new followers to your Instagram account. But properly networking with the right people is also going to lead to more followers, meaning more people will be seeing the images you post, and the product lines that you offer for sale.

See who follows you. If they are an established follower and if they always like photos you post and comment on photos, then follow them back. This is not only going to show your customers you care, it is also going to allow their friends and other followers to see who is following them back. In turn their followers will also follow you, meaning that by following one dedicated and established customer, you are going to gain several additional followers.

This simple networking trick will work wonders if an established follower is another business, or an individual who has thousands of followers themselves. More people will see your Instagram handle, will visit your page and more people will eventually follow you back as well.

Like and leave comments in return. When you are following others on Instagram, whether it is an established customer or another company, like their photos and leave comments on their photos. Just like you check to see who is liking and leaving comments on your photos and following them back, if people who you are following see you are doing this for them, they are generally going to follow you back.

You are gaining at least one new follower for your business simply by doing a little bit of networking, and engaging in a little conversation with other followers. In most instances, the person who you are following, will also have a list of dedicated followers. If they see you continually post comments on photos, they may ask for you to follow them and in turn will follow your account page as well. So you can gain several new followers simply by hitting the thumbs up, and by posting a few comments every once in a while to another Instagram account that your business is following.

Don't forget about the other social medias. Facebook has already been discussed, but Twitter and Google+ are also extremely important for your business as well. If you are using certain hashtags on these social media pages to describe new products or other important information about your business, make sure to include them on the new photos which you post to Instagram as well.

What this will do is it will automatically let your Twitter and Google+ followers view the hashtags; in turn, they will visit your Instagram page, and it may also lead to them following you on this social media platform as well. And if they share this link, or share the hashtag on their social media accounts, it is also going to lead other friends and followers of theirs to follow your business page as well.

Follow Brands

Of course your customers are important; but, other big companies and brands which are established on Instagram and have thousands, hundreds of thousands or even millions of followers can be just as beneficial to you.

So, find out which of these big companies or brands are following you (one such tool you can use to track which other companies are following your page is Statigram). Just enter the hashtag or the company name, and you can see if they have an Instagram page.

Why Other Brands Matter

You don't want to sit there and comment on every photo a competitor posts, but you do want to analyze what they are doing if they have an Instagram account. First off, you want to see what their followers (who are not your customers) appreciate in their brand, and not in yours.

Second, you can glean the images they post, how many likes they have, what is working and what is not working. And you can also get an idea of what your competitors are doing, when they post new product lines or up and coming ideas they have.

All of this is important for your business. Not only does it give you a heads up as to what your competitors are doing, it also gives you a way to potentially draw their followers away from them to following you. When you are aware of what the competition is up to, you are going to be on top of your game, and are going to make the right moves so you can keep up with the changes in the industry.
If you are not following a competitor, or if you do not know what they are doing, make sure you immediately do this. It will give you a great deal of insight as to how they operate, what they might have or might do that you are not doing and will show you what changes should be made. This will ensure you are appealing to the right audience, and can potentially draw in new followers to your account by making the right changes and updates to your own Instagram business page that you have set up.

Networking is key in any industry; just because the internet has simplified how we connect, and how customers view us, does not mean we can stop networking as a business. Not only does networking help you meet more people that can help you, it is an easy way to meet new customers and prospective customers who may never have known that your business existed.

As a business owner you have to network through Instagram, as well as the other social media platforms, just as you would if you were attending a local networking or social gathering event in the industry. It is not only simple for you to do, it is an easy way for you to glean what your competitors are up to, what customers in your industry are looking for and what changes have been made in the industry that you are not yet up to date with.

By using these simple techniques to network online, your business is not only going to see a jump in the number of followers on Instagram, but possibly on other social media platforms that you have set up as a business as well. When your followers see you care, and are thankful that they follow you, they will tell others about your business which will lead to new followers. And when you are sharing your information and accounts through all social media platforms, it allows current followers who were not aware of your Instagram account to follow you on that platform as well.

It is important to network and to utilize all of the tools that are available to you as a business owner. This will ensure you are gaining followers and growing through social media. These are a few simple tools you can use in any industry, to see growth and to realize new followers to your photo sharing page.

This will in turn lead to more new customers, more site visitors to your online business page and eventually to more profits due to the increase in the number of customers you are gaining through social media.

CHAPTER 7:
USE INSTAGRAM VIDEOS TO SHOWCASE BRAND MILESTONES

Images are excellent. They allow your followers to see the product, see the development and to see what goes on behind the scenes. However, images can only go so far; just like words they are limited in that they do not move, nor do they provide sound.

Although images are a step up from simply having lines of text explaining things, the addition of video, such as videos to your Instagram account is something that you should consider integrating to your business account as well. It gives your followers yet one more way to see what is going on behind the scenes, what product lines you are going to come up with next and other fun snippets that you choose to share through your account.

Just like Twitter introduced Vine videos, Instagram has come up with its own video format, allowing you to share short video clips just as you would share images on your account.

What Can You Do with Video?

There are a number of benefits that Instagram videos provide to business customers.

Some of the benefits of using video include:

- You can create fifteen second clips, in comparison to only 6.5 seconds allowed for Vine videos on Twitter. This gives your followers more to see, more to enjoy and a longer glimpse of wheat's to come in the future.

- You can use the same filters as you would use on your images. You can still use the great tools that Instagram has to offer with image posting to create the best quality scripts possible when you choose to share video clips with your followers.

- You can edit your videos. If there is a problem with volume, if something isn't in tune with the audio or if there are other issues with your video, you can use editing tools to correct these errors.

With Instagram video not only do your followers get to see fun clips you choose to share, they also get to see them in the best resolution and in the best lighting possible as well. When you post your videos, make sure to make use of these tools just as you would when filtering an image that you post to the site. It will allow your followers to see the video in the best resolution and quality, and will allow them to truly appreciate the short clip that you have put together to share and post on your account.

Instagram vs Vine

Creation		
Video length	15 seconds	6 seconds
Filters	Yes	No
Delete Last Clip	Yes	No
Import videos	No	No
Front-facing camera	Yes	Yes
Image stabilization	Yes	No
Drafts	No	(Probably) coming soon
Save to Camera Roll	Yes	Yes
Consumption		
Cover frame	Yes	No
Autoplay	Yes	Yes
Automatic sound	Yes	Yes
Share to	Facebook, Twitter, Tumblr, Flickr, Email, foursquare	Facebook, Twitter
Looping	No	Yes
Embeddable	No	Yes
Geotagging	Yes	Yes
Photo map	Yes	No

"Instagram vs Vine" by Jordan Crooks

Above are a few of the additional differences between Vine and Instagram video. Not only can you do and share more, you can save the videos and you can create high quality content, which is always what you want to share with your followers through your business account.

Don't Stop with Instagram!

If you have a blog, a business site or other online pages, your followers can read about your products and services.

Why not embed the videos you post on Instagram to these sites as well? Not only is it going to elicit more traffic, it is going to allow followers of different social media platforms and of your blogs to view the videos as well. So even if they do not have an Instagram account, they are going to be able to view the video that you have shared with other followers.

It is important to know when and how to use images and videos, and when to embed them. Do not overuse this feature, as it can create a "disconnect". Some people who are on your blog are there for the simplicity, and because they like reading about updates and changes. Make sure to keep this in mind, and only embed the videos when it is relevant, when it is necessary, when it is going to add some kind of value to a blog post or when you are posting a new story on any one of your other online sites.

By embedding, you are going to extend the reach of your videos, and are going to expand on the number of people who will view your video. Using Instagram alone, you never know who will share the video, or how many of your followers are actually going to see every video post that you choose to share on your account.

By simply embedding the video on to other sites, such as a blog, you increase visibility and the amount of times the video post is going to be seen. It will help to extend the reach of the content you are creating, extend the number of times the content is going to be shared and will expand upon the potential audience.

When Should You Use Video?

There are so many things that video can convey to an audience that an image just won't convey.

If your company is coming out with a new revolutionary product, you can share this in video. If you are reaching a milestone, whether it is the number of clients you have or the number of years you have been in business, video can convey this extremely well. If you would like to hold a competition, or would like to ask your customers to create content for your site, using video is the ideal medium to go through in order to connect your customers and the business in one location.

Using video simply to show an employee being silly at work won't have an impact on your followers; in fact, it might have a negative impact if the video does not add any value to them. But, if you are discussing new products which are going to be released, upcoming sale events, closeout deals or other information that video content will convey to them, in an enjoyable and easy to understand manner, then why not use it to reach out to your followers?

Fan appreciation videos are even great, as they show your customers you are thankful that they have chosen to do business with you, and you can always offer them some kind of incentive to continue doing business with you in the future through these short video clips.

Include Your Followers in Milestones

On a daily basis, businesses can reach a big milestone.

If a new company has just earned one thousand followers, this might be a great accomplishment. If the business has been in business for fifty years, this is a milestone that not every business is going to reach in its lifetime.

Why not share these, and other milestones with your followers? By making them feel as if they are a part of your success, they are going to want to continue being a customer of yours, and continue doing business with your company.

The minute your followers and customers feel unappreciated, they are going to turn elsewhere, and are possibly going to turn to one of your competitors. A simple way to not allow this to happen, and to keep your customers happy, is to keep them in the loop and to always make them feel as if they are a part of the business. You want your followers to feel they are a major reason that your business has been such as major success for so many years.

With the addition of video to Instagram, you can easily do this for your customers. You can share the major successes, you can include them in major changes that are going to come in the near future and you can even interact by asking your followers and customers for suggestions as to how you can improve your business and the quality of service you offer to them.

The milestones, accomplishments and big news you want to share with your followers will be far more significant, and will have far more of an impact on them when it comes in video form. This is opposed to simply posting a photo that says 50 year anniversary on it.

Personalize the Videos

Using video is a great way to personalize a message as well. It allows your followers to feel as if they are a part of the team, and that you want to include them in up and coming decisions to be made, or new product innovations that you have in line for the company. Regardless of what it is that you choose to share via video, it is a great way to connect with your followers and show they you truly care about them.

Video gives viewers the impression that you are actually speaking to them, asking for their opinion or otherwise would like to gain more knowledge as to what they are thinking about something that you are discussing in that video. For this reason, it is important to create messages which connect you to the viewer, which allow your followers to feel they are a part of the conversation or discussion and to create content that is interesting. You want to make sure you can keep them engaged for the entire fifteen seconds that they watch the video.

The Impact of Video

Video is an easy way to bring things to life. The things that your followers usually see on several images are going to come to life. When they see a celebration, they are going to see the people behind it.

No matter what you choose to use the video posts for, make sure it is something that is relevant to your followers and customers, and something that includes them as a part of the family that you have built. The impact a short video can have on a customer can help retain them as a customer and draw in several new customers, or can draw them to your competitors. So, make sure your video posts are properly utilized, and that they are done in tasteful manner when you choose to share them on Instagram with your followers.

Using video is simply one more way to connect with followers, and give them a little more insight as to what goes on within the organization. Using video will allow you to showcase major accomplishments, changes or new things which are going to come in the future for your company.

Additionally, the use of video is going to allow you to connect with your customers and your Instagram followers in a way an image doesn't allow you to do. So using video is a great way to connect, keep your followers in the loop and to make them feel as if they are an integral part of the business you have built over the years.

CHAPTER 8:
CREATE A VIRAL TREND BY SHOWCASING YOUR BRAND

Making an Instagram post go viral is one of the best ways to get your brand and business name out there. One viral post can turn your world around as a business owner, can lead to new business opportunities new followers and customers and of course far higher profit margins over time.

Making a post that will go viral on your Instagram account is not as easy as it would be through other social media and video sharing sites. For instances, using the right words, descriptive title and the right screen shot when you post an image on YouTube might lead to millions of views in a week's time.

With Instagram videos this is not the case. However, it is not impossible for your videos that you post through Instagram to go viral, so long as they are properly uploaded, properly shared, and seen on as many outlets as possible. If you do choose to use videos on your Instagram account, and choose to share this content with your followers, videos that do go viral are going to do wonders for your business and are going to lead in an increase of Instagram followers in no time as well.

If it is difficult for video content to go viral, just imagine how much harder it will be for images you share on Instagram to go viral. It is important to create an experience, showcase your brand in a way no other company or competitor can compete with and to come up with ideas that are going to go well beyond the imagination of your followers.

Although it is difficult, it is possible. With the right planning, the right look and the right approach to posting video and images on Instagram, you can create content that will go viral and that will help you gain that online recognition. This will draw in more followers to your account, and will eventually lead to the increased sale figures that you want to see as a business owner.

Create an Experience

One such company that has used Instagram to their benefit in terms of their posts going viral, is Sharpie. The company uses their markers and other products to show their customer how to create something fun. From drawing images on a canvas, to creating a whole new world with a few different colored markers, most of the images that are posted on the company's feed use a variety of colors, show images popping up off the screen and show some form of creativity flowing off the page.

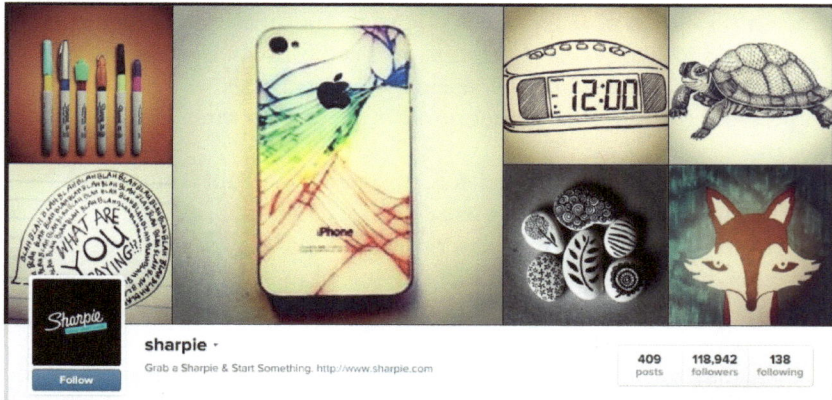

Sharpie Instagram Profile, @sharpie, September, 2014 by sharpie on Instagram

Why do this? First off, it gives your followers something fun to look at and shows them that the possibilities are endless when they are using your products. Whether they are working on a school project or a new idea to present to a client, with the product they are using they are going to be able to complete any task.

Using creativity is also a great way to increase discussion. The more people who like a video or an image, the more comments that are left in a feed and the more people who see such feeds, the more likely the chances are that these images or videos are going to go viral.

Something that goes viral is pretty much something that garners conversation; whether that conversation is positive or negative, it is good for a business. It allows people to hear the company name, it will force people who are not familiar with the company to go look it up and learn more about it and it will even draw people to visit a company site.

This is the end goal for all companies who are using Instagram and other social media sites. So why not incorporate a bit of creativity, even if it results in conversations? As long as people are talking about the business and as long as they see different possibilities with the business, then there is something interesting that your business can offer (which another business can't offer).

Let Creativity Lead the Way

No matter what the product or service is that you sell, when posting content on your Instagram account and other social media outlets, businesses should allow their imagination to run rampant. Imagine creating something that your customer could see themselves using. Whether it is a sharpie or any other product, show the customer how that product is going to better their lives, or how it is going to allow them to become a happier person if they own it.

If you can't think of something creative and unique about your own product, how can you expect your customers to do so? You are the creator of the product, therefore you have to show customers some of the countless ways it is that your product is going to improve their life, is going to give them something they do not have and is going to give them something that they are not going to be able to get from any other company or product that is available on the market.

Make It Unbelievable

Content that goes viral is generally content that people can't believe they are seeing on the screen. It is something they would never post themselves, or it is content that has never been seen before. Just imagine the early videos that went viral; animals doing crazy things, babies dancing and other content that had never been shared in this past.

Being creative is a part of thinking outside the box, and coming up with something that has never been done in the past. Remember, a viral video can do as much harm to your business as it can do well. So when creating content that is unbelievable, make sure to avoid putting other companies down (and definitely DO NOT insult their customers), or to potentially segregate a certain demographic group.

You want the content to be fun, you want it to be unique and you want it to have that "wow factor". You have to remember that too much of anything, even if it is fun and good, can be viewed as a negative. In creating your content, make sure it has that "wow factor", make sure it gets people talking and make sure it shows people how much fun and how they can create with your product.

Showcase Your Company as a Leader

Again, the videos and images that are going to go viral are the ones which have never been seen. The images that are unique, creative and out there, will create something that viewers have never seen in the past. You want to showcase your business as an innovator, as a company that is a leader and as a company that is going to take charge in the industry that it is in.

Your industry has to be beloved, if you want visitors to go to your site and to view your content. The more people like the company, the brand and what it stands for, the more likely it is they are going to appreciate the content.

Even if it is a product a customer does not need, or something they are not currently looking for, if you are a leader, an innovator and if the customer loves you, it is going to lead to more viral content and eventually more people seeing your site and products that your company offers for sale.

Include Customers

Sharpie takes their followers on a ride. With each image they draw (which goes viral), their Instagram followers actually see the hand that is drawing the image. They see the image come to life, and they take the audience on a journey with them. If the customer does not feel as if they are a part of the journey, no matter how creative it is, they are not going to get on board with the product you are trying to sell to them.

So, in creating the videos for Instagram and in deciding which images you are going to post to your page, make sure your audience feels as if they are a part of the product, of the journey and of the content that you are sharing with them.

If a video or image feels forced, or as if it is being shared simply for the company's self-gain, followers are not going to get on board. You want people to like and comment on content so that it is shared, and so it eventually goes viral. The only way for people to do this, is for them to feel as if the product you are selling has something to offer to them. Make your content engaging, and think about your posts through your followers' eyes to ensure you are posting content that they are going to connect with.

Instagram is a great platform to help your business grow; it gives you more ways to share, and gives you the visual ways to share content which other social media platforms did not offer in the past. As a business owner, you have to take advantage of this if you want your business to grow. However, creating content that will eventually go viral is far more difficult on Instagram than it is on other platforms and video sharing sites. It is not impossible to do; in fact, many major companies have a number of images and videos which they post (on a daily basis), that go viral.

When content goes viral, it means it is going to be seen by millions, possibly even by people who have no idea what Instagram or any other social media site is. If you want your content to go viral, make sure you use the right tactics. Keep it playful, always sway towards the creative side and keep your followers in mind. If you are simply doing things for the potential of business gains, this is going to show in your content. When you create the content with the intention of being fun for your followers and customers, and to showcase how amazing your product really is, this is what will help push images and videos towards going viral.

CHAPTER 9:
USE FACEBOOK TO HELP YOUR PHOTO CONTESTS ON INSTAGRAM

Photo contests are also a great way to involve your followers on Instagram and to engage with your business online as well. Using Facebook is an easy way to integrate RSS feeds and hashtags so your followers know what to include when they share a photo through Facebook.

Not only will more people see the photos, but photo sharing competitions are also a good way to give those who take part in the competition a great prize, to show them your appreciation for them taking part in the contest and helping to spread the word to other followers as well.

Host a Campaign

Instagram can be used to host a photo contest of campaign. It is quite simple for you to do through your account page. The use of hashtags is the most important factor when hosting a campaign; it is the way you are going to collect the images which are submitted by your followers, so you can properly categorize them once they are received.

Users will be able to click on a tag. One example is a company hosting a competition for the best shoe design. When posting an image, the contestant simply has to add the #shoe (or other hashtag), and their image will be tagged to that word or phrase. From there, other users can click on the tag, will see the latest submissions and will see the submissions with the most comments.

It is important to create unique hashtags so that your followers know which one to use when posting an image, and so the images they share for the contest will be properly categorized when the judging takes place.

Use RSS Feeds

When hosting these competitions, each hashtag that you use is going to have an RSS feed. This is what is going to make it possible for you to connect through Facebook, and through other devices which are capable of opening up an RSS feed. Whichever images the users are interested in, they will simply point their device at the RSS tag, and it will pull up that photo as well as every additional photo which has the same tag.

Using the RSS feeds is the best way for event organizers to keep track of all the images that are submitted. It will allow them to categorize the feeds on Facebook as well as on Instagram, so that users will quickly and easily be able to open an image or a specific category when they sign on.

In addition to the event organizers, anyone who wants to see the images which have been submitted are going to be able to do so simply by pointing the RSS reader device at the RSS tag, which will open the posts that are linked to that specific tag.

How to Promote On Facebook

If you want to promote the contest on Facebook, the easiest way to do so is to come up with a creative hashtag, and one that is going to distinguish the competition. Not only will allow Facebook followers have the ability to see the images, see the competition and learn about the details, but they will also be able to post their own images through Facebook or through Instagram when they use the right hashtag with their image post.

By doing this, not only will a competition gain more followers and more interest, it is also going to elicit more people to want to engage in the contest and to post their own submissions, which are going to help you as the business owner in the end.

Therefore, integrating Facebook in to the mix, and using the appropriate hashtags, is a must if you want your contest to have as many people as possible taking place in it and as many images as possible being submitted to the contest.

Use the Right Hashtag

Regardless of what you want your followers to submit, or what type of image you want them to come up with, using the appropriate hashtag will make a world of difference. Creating a hashtag that is memorable, easy to remember, is not difficult to submit or write and one that people can easily tag are a few factors to keep in mind when holding such competitions.

Not only will it be easier for people to submit their photo, it will also be easier to categorize different images for the RSS feed when they are going to be shared on Facebook or other social media platforms by your business.

Again, keep in mind that hashtags should be business-related. If it has nothing to do with your industry, or if it is something that your followers are going to have a hard time remembering, it is highly likely they are not going to engage in the contest, meaning you will receive fewer entries than you had hoped for. By making the hashtag industry-specific, it allows other people who might not be following your business online, to learn about the contest and to take part in it as well.

Leverage Using Facebook

Once you have posted the contest information, in order to elicit as many images as possible from your followers, you want to keep them in the loop.

A good way to do this is to promote using Facebook. You can inform them of the prizes that are going to be given out to winners, or let them know what the images are going to be used for once the contest is over. You want to give people incentive to take part in your contest; after all, you are going to use the images that they submit, and the creative message that is submitted along with the hashtag for your business benefit.

Since this is the case, people who take part are going to expect something in return. Regardless of how big or small it is, make sure people understand what the contest is about, what they are going to receive if they win (if anything) and every other piece of relevant information about the contest.

By promoting the contest on Facebook, you are going to elicit more people to take part in it. More people are going to view it, and more people are going to share it. If a follower likes the contest, this will automatically inform their friends on Facebook of the contest as well. In turn, more people are going to learn about the contest when you use the social media site to promote it, and you are going to receive more response and more photos for the contest when a larger audience is aware of it.

Using the social media platform is in your best interest and it does not take much effort for you to set it up.

More Likes and Views

Using Facebook to promote your contest is also going to result in more comments, more likes and more views. What this means is that more people are going to know about the contest, even the ones who are not following you on Instagram.

If they want to take part in the contest, they will then follow your business on Instagram, meaning you are going to gain new followers. Plus you are going to use the visibility of Facebook to continually increase the number of participants that are going to send in photos for the contest that you are hosting.

Use Facebook Status Updates

Status updates on Facebook are a great way to keep people in the loop as to what is going on. You can inform your followers how many days they have left to submit an image, or if the contest is going to close early. If you plan on extending the contest due to the number of applicants and images that were received, you can also take to your Facebook feed to let your followers know about this. This will ensure you continue receiving images from them when hosting a photo contest.

No matter what you want your followers to know, what new hashtag you want to incorporate or what other changes you want to make to the contest, you can always use your Facebook feed to keep them in the loop. Not only are instant updates sent to email and to mobile devices, but more people are likely to see the changes and updates as they take place when they are linked to your social feeds on Facebook.

Engage Your Followers to Join In

One of the main reasons to host a photo sharing contest is to engage your followers. If you are going to create a new product, ask them to send in images for you to promote it online. Not only are you getting free design work, you are going to make your customers feel like they are a part of the process, and like you truly want them to enjoy the new product line that you are planning on introducing to them in the future.

By engaging your followers, by keeping them in the loop and by allowing them to take part in the design or implementation phases, it will bring you and your customers closer. Even if you are not giving out a physical prize in the contest, the simple fact that you want their input and want them to help in the design phases, is going to be incentive enough for hundreds if not thousands of current (and new) followers to engage in the contest.

Although Instagram is a great place to share images, if you are planning on getting your customers and followers involved, using the power of Facebook is a great way to increase the number of participants and to ensure as many people as possible take part in your social competitions. Facebook allows you to connect to Instagram, allows more people to see the feeds and updates and allows people who are not following you on Instagram, to take part in the contest.

The end result is that you will receive more photos, your contest will garner more attention and you will eventually receive the best content to use in the end. So, use the feeds, promote, and keep your followers in the loop using Facebook to leverage your Instagram contests.

CHAPTER 10:
REWARD YOUR FOLLOWERS WITH PROMOTIONS AND FREEBIES

As a business owner, you want people to follow you on Instagram and through other social media platforms. The main reason that you want this is for them to make purchases from you, and to potentially give your company referral business by telling their family and friends about the great products that you offer for sale.

What exactly are you giving your customers in return when they follow you? Since you want something from them, it is important that you show your gratitude as a business owner when they do follow you through social media sites, tell people about your great business and make purchases from your company.

There are a number of ways you can reward your customers and followers, not only for following you, but for referral business and for their loyalty. With so many competitors that they could turn to in nearly any market or industry that your business is in, they decided to follow your company.

One company that is most well-known for these special promos is American Express. They have been known to give away first class tickets, backstage concert tickets or even tickets to sporting events which are impossible to find (such as the US Open).

American Express Instagram Profile, @americanexpress, September, 2014, by American Express

More often than not, the company gets these deals for free or a very low cost as the promos they offer are through their partners. Simply offering these promotions or special items to their online followers is a great way to thank them for their loyalty and to ask them to continue using their products, as opposed to turning to the competition.

These are a few ways to keep your customers coming back, keep them following you on social media and to keep them from turning to competitors in the same industry.

Offer Followers Promo Codes or Sales

One simple way to give back to your followers is through promo codes or other special sales. You can offer your Instagram followers a special promo code which can only be seen by those who are following the company through the social media platform. Or, you can inform your followers online, about sales that are going to take place before the general public learns about the offers.

Doing these simple but personal things for your loyal followers is a great way to thank them for their business. You were going to host the sale anyway, why not inform the online followers a few hours or even days before, so they can take advantage of the items they want to buy?

Or, why not give them certain online promo codes, which can only be used by those who follow them through certain social media platforms? When you do this, you give people an incentive to continue to follow you and to check in on social media platforms regularly.

Offer Incentives for Comments

Another simple way to grow in the number of likes, comments and followers is to offer incentives for these simple tasks. You can promote a contest where if new customers follow you through the social media site, they are automatically entered to win tickets to a big event.

Or, if a follower on Instagram likes or comments on a photo, they are entered to win a ticket or a product that has not been released for sale by the company yet.

Although these are extremely simple tasks for people to do, a number of people do not do them simply because they have no incentive to do so. When you receive likes on a photo, or when there are hundreds of comments on a photo, it has more of a chance of going viral which is exactly what you want as a business owner.
A great way to get customers and potential customers to help you in taking your site and social media pages viral, is to offer them something in return for helping you. Not only will more people share your site and your content, but the incentive does not even have to be that large for a majority of the people to take part in the contest, or to share a photo that your business posts on their Instagram account page.

Offer Rewards for Following

If you are an airline company, offer customers reward points for travel if they share your photo on Instagram. Or, if your company sells clothing, offer a customer a coupon for $15 if they share something that you posted on your page. These are simple ways to reward your followers, to keep them happy and to avoid losing them to other competitors out there.

Most companies offer their followers at least some form of a reward; even if it is something small, it is a form of gratitude that you are thankful they are your customer and that they are following you on social media platforms.

If your business does not offer any form of an incentive, or if you do not offer any reward to your followers, it is going to result in them turning to the competition. Again, you want these people to keep following you, and you want them to remain loyal to your company. But, why would they do so if you do not give them anything in return, and if you do not show your appreciation for them being a loyal customer?

As a business owner, you have to offer your customers who are following you on social media some form of reward. Whether it is quarterly, once a year, twice a year or even in the form of a contest, it is a small way to show them you are thankful that they are following your site, and that you want them to continue following you on as many social media platforms as possible.

The more rewards that are offered, and the more opportunities your followers have to earn rewards, the greater the chances are they will continue to follow you through social media platforms.

Give Followers Free Perks

You can offer things such as free trips, gift cards or credit card gift rewards, movie tickets, backstage concert tickets or tickets to their favorite sporting event of the year. These are a handful of perks that companies have offered in the past, and continue to offer to their followers in order to keep them on board.

As a customer, you want to be shown some form of gratification. Not only are you following the company on social media, you are taking the time to like and comment on their videos and images. Furthermore, you are spending money when you shop through their online store, or when you shop with one of their local retailers. You are doing so much for the business, so you naturally want something in return for your loyalty and for the spending habits that you have formed over the years with the company.

The companies that offer these perks and give their customers something in return, are not only the ones who have the most followers on their social media platforms, but are also the businesses which are most well-known around the world. These are the businesses that tend to earn the highest profit margins, and will sell nearly any product to their customers. The customers will purchase these products because they rely on the company.

In order for your business to survive, it is essential for you to give back to your customers. Like any other relationship, you can't expect one side to do all of the work. If your customers are continually doing things for you, and to help your business grow, it is important that you do something in return for them.

So, offer them perks a few times a year, or give them the opportunity to win a cool prize. Even if they do not win, it is going to show them you care about them, it is going to show your customers you are thankful for their loyalty and it will show them you want to retain them as a customer and keep doing business with them in the future.

Tit-for-Tat

If you have social media platforms for your business, you will want them to grow. Even if you choose not to give away from concert tickets, or give away money for people following you, you can offer contests and other forms of incentives for them to offer to you.

Not every business has as much money to spare as American Express; this means not all businesses are going to be able to give away trips and expensive prizes to their followers. But, you have to show some effort, and you have to show your followers you are willing to give them something in return if they follow you on social media platforms.

It can be as simple as giving them a few points towards a future purchase when they comment on a photo you post. Or, you can offer them a 10% discount at checkout when they shop online (or free shipping) if they like a post you have made on your social media pages.

It does not have to be a grand gesture, but there has to be some kind of gesture. If you want people to follow you and to take the time to comment on your social media platforms, you have to give them a reason to want to and some form of incentive for them to do so.

No matter what line of business you are in or what social media pages you have, the more followers you have online, the greater your chances are of growing online and growing at local retail shops you operate. If you are a brand new company, or even a well-developed company and are just getting in to the social media scene, make sure you understand how it works. It is important to learn from the companies that have millions of followers. Look at what they do, and how they are thankful to the followers they have on the social media pages which they operate.

By showing your followers some form of gratitude and offering them an incentive to follow you, it is not only going to make them want to keep following you, but to actually engage (by commenting, liking or even taking part in contest) on your social media pages. So remember to give them returns, to reward your followers and to continually thank them for remaining loyal to your business, even though they have other choice competitors to turn to.

ABOUT THE AUTHOR

Issa Asad is an entrepreneur and marketing strategist from Miami, Florida. Mr. Asad has over 15 years of experience in the marketing, technology and telecommunications industries. Mr. Asad has been the Co-Founder, Founder, Managing Member, President, Vice President and Chief Executive Officer (CEO) of multiple technology, telecom and online-based companies.

Currently, Mr. Asad is the CEO of Quadrant Holdings, LLC, a holding company located in Dania, Florida. He is also the CEO of Centurion Logistics, LLC, a technology and customer relation management developer that provides transaction processing to Continuity Marketing businesses.

www.ingramcontent.com/pod-product-compliance
Lightning Source LLC
Chambersburg PA
CBHW041717200326
41520CB00001B/134

www.ingramcontent.com/pod-product-compliance
Lightning Source LLC
Chambersburg PA
CBHW041717200326
41520CB00001B/134